YOUR KNOWLEDGE HAS VALUE

- We will publish your bachelor's and master's thesis, essays and papers

- Your own eBook and book - sold worldwide in all relevant shops

- Earn money with each sale

Upload your text at www.GRIN.com and publish for free

Production of extruded meat substitutes based on textured soy protein

Naomi Albiez

Bibliographic information published by the German National Library:

The German National Library lists this publication in the National Bibliography; detailed bibliographic data are available on the Internet at http://dnb.dnb.de.

ISBN: 9783346641137
This book is also available as an ebook.

© GRIN Publishing GmbH
Nymphenburger Straße 86
80636 München

All rights reserved

Print and binding: Books on Demand GmbH, Norderstedt, Germany
Printed on acid-free paper from responsible sources.

The present work has been carefully prepared. Nevertheless, authors and publishers do not incur liability for the correctness of information, notes, links and advice as well as any printing errors.

GRIN web shop: https://www.grin.com/document/1193250

Faculty of Horticulture and Food Technology

Project Thesis

Production of extruded meat substitutes based on textured soy protein

Author: Naomi Albiez

15. Dezember 2021

Module: Food Biotechnology

TABLE OF CONTENTS

1. Introduction and aim of the study 1

2. Soy protein as meat substitute 1

   2.1. Structure and function 2

   2.2. Protein fractions.. 4

   2.3. Advantages of soy protein in extrusion... 5

3. Manufacturing processes of protein isolates, concentrates and flours.................. 6

4. Extrusion process 8

   4.1. Basics of extrusion 9

   4.2. Extruder construction.. 10

5. Conclusion and further considerations.. 12

6. References... 14

7. Figures 16

8. Tables 16

1. Introduction and aim of the study

Due to the steadily growing world population and the rising wealth in developing countries, the demand for proteins is increasing (Vázquez-Rowe 2020). However, it is difficult to cover the demand for proteins only with animal proteins. Furthermore, various customers attach great importance to their well-being and health, and therefore avoid eating meat for ecological, ethical and social reasons. Hence, it is of great interest to access domestic plants with protein-rich seeds for the market of meat alternatives.

On the food industry side, a trend towards the production of plant-based, protein-rich products for human consumption is emerging. In the first quarter of 2020, sales of meat substitutes in Germany increased by 37% compared to the first quarter of 2019 from just under 14.7 thousand tons to 20 thousand tons (Statistisches Bundesamt 2020).

To make consumers choose the plant-based alternative, these products need to imitate the techno-functional and sensory properties of animal products. In the 1960s, the first meat analogues using an extruder were produced from common proteins like soybean protein (Osen et al. 2014). However, extrusion is a very complex process in which numerous parameters interact with each other. The effects of different material, machine and process parameters on the physical properties of the end product are very difficult to predict. For this reason, food extrusion is still carried out according to the "trial-and-error" principle (Schuchmann and Danner 2000). Among other things, this leads to a high loss of raw materials and efficiency.

The aim of this work is therefore to provide the fundamentals of extrusion. This includes an in-depth understanding of the structure and composition of plant proteins. Based on this, the production of the raw materials is discussed, as this has a great influence on the final product. In the last step, the difference between high moisture extrusion and low moisture extrusion is explained, as well as the general set-up of an extruder.

2. Soy protein as meat substitute

Along with carbohydrates and fat, proteins make up a large part of our diet as they provide the necessary building blocks for protein biosynthesis. Plant proteins are often used as the basis for meat substitutes. Sales are not only driven by environmental and ethical considerations, but also by health effects. The World Health Organization

(WHO) has classified red meat as potentially carcinogenic to humans. Processed meat has been linked to diabetes or diseases of the cardiovascular system. A general shift to a plant-based diet high in legumes and whole grains could reduce the global mortality rates by 6 - 10%. (Godfray et al., 2018)

Meat and fish substitutes can help to enable this transition. These products usually contain so-called textured vegetable protein (TVP) and have a protein content of 50 - 95%, depending on the source. Commonly, such products are made from soy, wheat or pea, but there also exist other native grains, oilseeds or legumes that can provide a rich source of protein. (Belitz, 2008)

Besides contributing to flavor and color, proteins provide important techno-functional properties promoting the product quality of food applications. They are able to stabilize gels or foams and form fibrillar structures. To enhance the above-mentioned physical properties and increase the nutritional content, proteins can be applied as flours, concentrates or isolates. In this way, they are used in the production of fish or meat-like foods. (Belitz, 2008)

2.1. STRUCTURE AND FUNCTION

Proteins, as macromolecules, play a central role in the functionality of foods and in biological systems. They are composed of amino acids linked by peptide bonds. The sequence of the amino acids determines the unique three-dimensional structure of each protein and its specific function (Khan, Siddiqi, & Salahuddin, 2017). Above a certain molecular weight, or more than 100 amino acids, it is no longer called a polypeptide but a protein (Vasudevan, Sreekumari, & Vaidyanathan, 2011).

Protein molecules take over all important structural and functional work in the body. In addition, from a nutritional point of view, there are two further reasons for the relevance of proteins. On the one hand, a varied diet covers the need for essential amino acids that cannot be synthesized by the human organism itself. For example, a high concentration of vital lysine is found in legumes. On the other hand, the supply of nitrogen ensures the formation of non-essential amino acids. Other nitrogen-containing compounds, such as nucleic acids and creatine, also depend on a regular supply of nitrogen. (Young & Pellet, 1994)

The side groups of the individual amino acids are decisive for the multitude of protein structures with the most diverse properties and functions. These are responsible for the three-dimensional molecular structure of the proteins through various interactions with each other. The basis is the so-called primary structure, thus the linear sequence of amino acids and the position of existing disulfide bonds. (Khan et al., 2017)

The primary structure is followed by the secondary structure. This is the spatial structure of the amino acid chain, which arises as a result of hydrogen bonds between neighboring functional groups. A distinction is made between the spiral α-helix and β-helical structures. Helical structures are mainly formed by amino acids with no or small side chains. However, in the β-sheet structure, the β-strands fold over each other. (Whitford, 2005)

The tertiary structure refers to the three-dimensional conformation of the entire protein. In this type of structure, amino acids are far apart in linear sequence, but are close to each other in three-dimensional view. This three-dimensionality between the individual parts of the molecule is maintained by non-covalent bonds, such as hydrophobic interactions, electrostatic interactions, or van der Waals interactions. In this context, the tertiary structure in the native protein is always the most thermodynamically stable. (Khan et al. 2017; Vasudevan et al. 2011)

When two or more peptide chains assemble to form a protein, this is known as the quaternary structure. These subunits can be similar or different, resulting in homogeneous or heterogeneous quaternary structures, respectively. The hydrophobic bonds make the greatest contribution to stabilizing the protein molecule. The

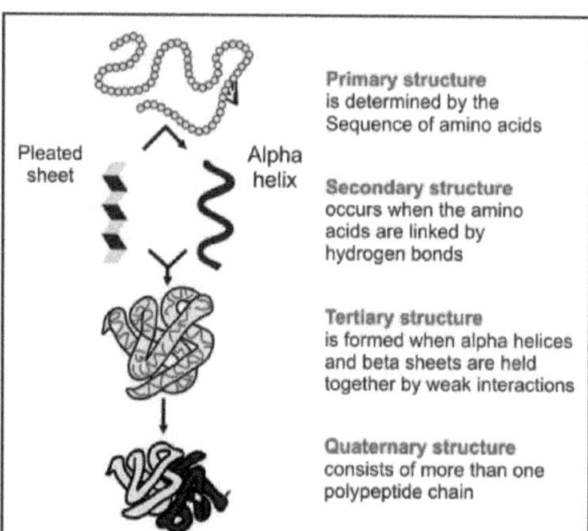

FIGURE 1: DIFFERENT LEVELS OF PROTEIN STRUCTURES (VASUDEVAN ET AL., 2011, P. 31)

interactions within the protein largely define the conformations and properties. (Khan et al. 2017; Vasudevan et al. 2011) The preceding explanations of the various protein structures are depicted in Figure 1.

During the extrusion process, the bonds are broken by shear and high temperatures, resulting in a change in the tertiary and quaternary structure. During this denaturation, new bonds can be formed between the proteins as side groups come to the surface and crosslink with each other. Partial unfolding and stretching of the globular proteins result in an arrangement of the protein strands in the direction of flow. This leads to new functional properties of the proteins. (Kristiawan et al. 2018; Belitz 2008)

2.2. PROTEIN FRACTIONS

Proteins are divided into four Osborne fractions according to their solubility behavior. A distinction is made between water-soluble albumins, globulins, which can be extracted with a sodium chloride solution and alcohol-soluble prolamins. All the proteins remaining in the residue are classified as glutelins. The storage proteins albumins and globulins originate to a large extent from the cytoplasm or other subcellular fractions, whereas prolamins and glutelins are storage proteins. (Belitz, 2008)

However, it must be mentioned that the classification of proteins according to this scheme depends on the conditions of production or seed pretreatment as well as on the type of fractionation. (González-Pérez & Vereijken, 2007)

Fractions can be separated into their major components by chromatography or ultracentrifugation. They vary in molecular weight and have different isoelectric points (IEP). This leads to a different behavior under thermomechanical stress. The solubility of the proteins is lowest at the IEP since the same amount of positive and negative charges is present there. As a result of the lack of electrostatic repulsion, the proteins aggregate via intermolecular hydrophobic bonds, insofar as sufficient exposed hydrophobic groups are present. (Belitz, 2008) The molecular weights and IEP of soy protein are presented in Table 1.

TABLE 1: PROTEIN FRACTIONS IN SOY PROTEIN

Protein Fraction	IEP	Molecular Weight [kDa]
7S-Globulin (Vicilin)	4.4 – 4.6	~ 180
11S-Globulin (Legumin)	4.4 – 4.6	~ 350

The proteins of soy can be divided into three fractions: albumins, globulins and glutelins. Globulins represent the largest fraction, accounting for 70% of the total protein content (Boye et al., 2010). Two main globulins exist in all legume species, with few exceptions, and are classified based on their sedimentation coefficient as vicilin (~ 7S) and legumin (~ 11S) (Belitz, 2008; Freitas, Ferreira, & Teixeira, 2000). Legumin has a significantly higher molecular weight of 340 - 360 kDa than vicilin with 175 – 180 kDa (Swanson, 1990). These proteins generally have minimal solubility at the pH values between four and five (IEP) (Boye et al., 2010). In contrast, for example, Swanson specifies the isoelectric point of pea albumin at a pH of 6.0 (Swanson, 1990).

For basic extrusion research, it is generally important to know the IEP of the different protein fractions, since protein solubility analysis according to Morr must be performed to determine the protein binding types. The successful performance of this analysis is strongly dependent on the IEP. It is not recommended to choose the isoelectric point to perform protein solubility analysis. At this point, protein solubility is at its lowest due to increasing protein interactions, as the electrostatic forces are at their minimum and less water can interact with the protein molecules. Therefore, the protein cannot be completely dissolved in the buffer substance, which leads to a falsification of the results.

2.3. ADVANTAGES OF SOY PROTEIN IN EXTRUSION

In addition to extrudability, other factors must be considered for the suitability of a raw material for efficient application as a meat substitute. From an economic point of view, the market price and availability play a major role. However, the final product must also be accepted by the consumer. More and more value is placed on regionality, ecological cultivation methods and non-genetically modified plants.

For people who do not want to consume animal proteins for ethical, religious or ecological reasons, soy offers the possibility of becoming an important protein source. Soy protein convinces through its abundant worldwide availability, high nutritional quality and its diverse functional properties (Wang et al., 2004). It is an excellent source of protein and contain fiber, vitamins and minerals, while being low in sugar, sodium and fat (Kristiawan et al., 2018). Furthermore, there are many scientific studies on extrusion with soy protein and it is already widely used to make vegan substitutes without bringing a strong inherent taste like pea protein, for example.

Protein powders from soy can be divided into three different purity grades. A precise classification is made in the next chapter, but soy protein concentrate is available at lower cost than soy protein isolate due to its lower protein content and is therefore mostly involved as a main ingredient in a number of food applications (Cheftel, Kitagawa, & Quéguiner, 1992). However, soy faces an increasing consumer rejection due to its association with rainforest deforestation and the existence of several genetically modified varieties. In addition, its high allergenic potential excludes a certain group of consumers.

3. MANUFACTURING PROCESSES OF PROTEIN ISOLATES, CONCENTRATES AND FLOURS

The selection of the protein fractionation has a great influence on the final product. In general, a decision is drawn between three degrees of purification. A distinction is made between protein flours with less than 70% protein in dry matter (DM), protein concentrates with at least 70% protein in DM and protein isolates with at least 90% protein in DM (Eldridge 1982). In comparison to a protein concentrate with 3.5% fiber content, a protein isolate has a much lower fiber content of 0.2% (Belitz 2008). During extrusion, in some cases, higher fiber contents can have a negative effect because they block bonds between protein macromolecules (Riaz 2000).

When selecting a raw material, financial concerns also play an important role. Soy protein concentrates and isolates are more expensive than soy protein flours due to further extraction steps. They are rarely used as the exclusive ingredient to produce textured vegetable protein (TVP). However, adding concentrate or isolate can optimize the product by improving the water holding capacity and protein content of TVP. It should be noted for process parameters, that a higher fat content of the raw material requires an increased temperature and more shear energy due to lubrication effects. (Riaz 2000)

Protein preparations are mainly obtained from flaked and defatted flour, the residue of oil extraction. Isolates with a higher protein content are obtained by first extracting the soluble ingredients from the protein flour with water or diluted alkali (pH 8.0 – 9.0) and then re-precipitating the proteins from the extract by adjusting to pH to 4.0 – 5.0. Protein denaturation greatly affects the amount of protein that can be extracted in protein isolates. To produce soy protein concentrates two methods can be used: acid-wash and alcohol-wash. First, the flakes are soaked in water and acidified at pH 4.0 – 5.0 respectively soaked in 60% alcohol. This is followed by separation of soluble ingredients by centrifugation, washing and drying of the residue. Protein concentrates from an acid-wash have significantly better functional properties than those from an alcohol-wash because alcohol has a strong denaturing power. (Belitz 2008; Wang et al. 2004) A schematic overview of both processes for the production of soy protein concentrate can be seen in Figure 2 and Figure 3.

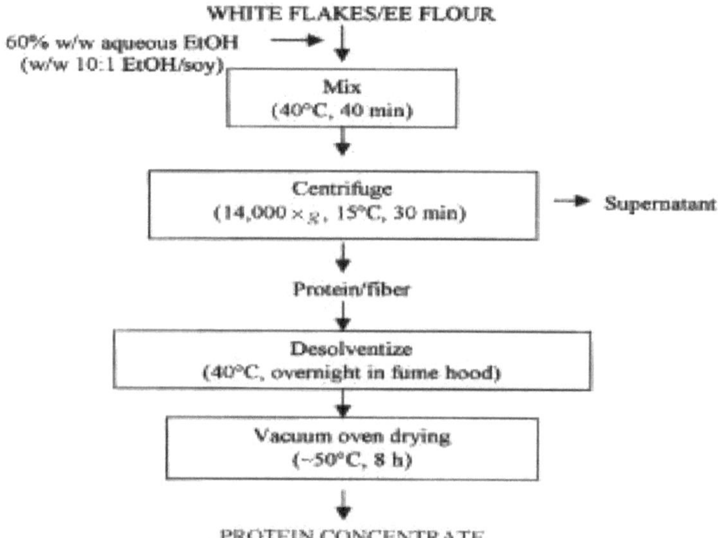

FIGURE 2: PROCESSES FOR THE PRODUCTION OF SOY PROTEIN CONCENTRATE VIA ALCOHOL-WASH METHOD(WANG ET AL. 2004)

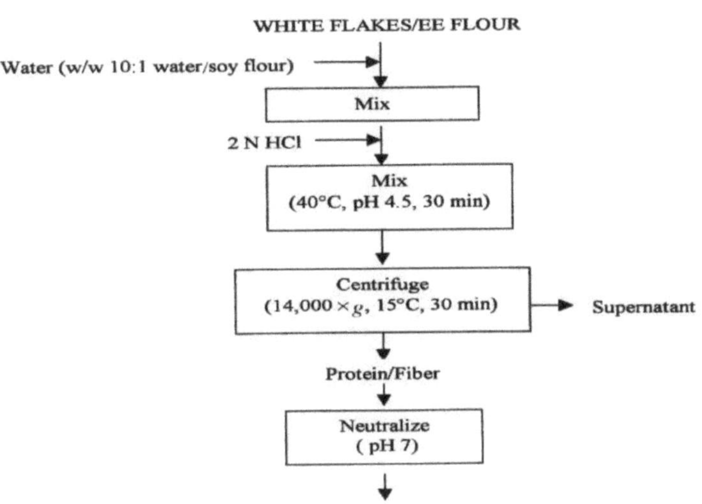

FIGURE 3: PROCESSES FOR THE PRODUCTION OF SOY PROTEIN CONCENTRATE VIA ACID-WASH METHOD (WANG ET AL. 2004)

This shows that the degree of denaturation of protein ingredients can be significantly influenced by the choice of extraction and protein isolation procedures (Mession et al. 2015; Liu and Hsieh 2007). Denaturation refers to the reversible or irreversible, partial or complete change of the native conformation of a protein. In this process, hydrogen bonds, ionic bonds or hydrophobic bonds are dissolved. (Belitz 2008) If the denaturation temperature of a protein is exceeded, the secondary, tertiary and quaternary structure can change. The proteins agglomerate. This results in a strong decrease in solubility and a change in the functional properties of the subsequent protein preparations. (Schuber 2008) In addition, the denaturation temperature of the raw materials during cooking extrusion determines the heat-induced gel network formation and thus has a decisive influence on the structure of the texturates (Shand et al. 2007).

4. EXTRUSION PROCESS

Extrusion is a continuous high-temperature and short-time process that is used to produce breakfast cereals, snacks, pasta or animal feed. A variety of protein-rich raw materials are particularly suitable to be converted with the extruder into modified intermediates or finished products, such as protein extrudates or meat analogs.

4.1. BASICS OF EXTRUSION

The different properties of the protein components, as well as the diverse extrusion conditions, lead to a complex process where the intercorrelating material, process, system and product parameters are still not fully understood (Osen et al. 2014). During extrusion, the materials undergo chemical and rheological changes due to a combination of high temperature, pressure, and shear, which strongly affects the texture quality of the extruded products (Osen et al. 2015). In the case of fish and meat analogs, the highest consumer acceptance is achieved through high-moisture extrusion, as the products produced in this way most closely resemble animal muscle fiber. (Liu and Hsieh 2007)

The previously common process is called low moisture extrusion (LME), in which doughs with moisture contents below 35% are processed. The molten mass leaves the extruder, water vapor escapes while pressure is released, leaving vacuoles between the branched protein strands (Belitz 2008). The result is a porous product that can then be ground into powder or granules (Guy 2001). Furthermore, LME products can be rehydrated by the consumer at home and used as a meat substitute (Gomes Arêas 1992; Guy 2001). However, they fail to imitate fibrous muscle meat (Osen et al. 2014).

Another method to obtain meat-like products from plant protein is high moisture extrusion (HME). The finished products are characterized by a high moisture content of over 50%. The combination of temperature, pressure, and shear leads to molecular transformation and chemical reaction of protein molecules, which contribute to the stabilization of the formed three-dimensional network (Liu and Hsieh 2007). A cooling die placed at the end prevents the evaporation of water during this process by sudden cooling the product below 100 °C (Wittek et al. 2020). This results in a fibrous, unexpanded structure (Osen et al. 2015).

For the industry, the extruder brings numerous advantages. First, it can be used in a variety of ways to manufacture products with different shapes and textures. In addition to a variety of other applications, the screw extruder is particularly suitable for the production of protein-rich products that imitate meat in structure (Wittek et al. 2020). Furthermore, extruders operate continuously and at high thermomechanical rates, so they produce food products in very short processing times. Productivity can be increased due to high throughput and full automation. It also means less space is required per unit of work compared to traditional cooking systems, resulting in lower manufacturing costs. Next, the optimal temperature-time ratio of the process must be mentioned as an advantage. Due to the high temperatures for a short time, fewer temperature-sensitive ingredients are damaged. At the same time, unwanted components such as enzymes and microorganisms are destroyed (Guy 2001; Harper 1981).

4.2. EXTRUDER CONSTRUCTION

Within one device proteins are plasticized, mixed, heated, sheared, and forced through a die to obtain the desired product properties and functionalities. During this process, the biopolymers are subjected to temperatures up to 200 °C and shear rates of up to 5000 s^{-1} (Wittek et al. 2020). Due to heating and shearing during the extrusion process, the macromolecules in the food ingredients lose their native structure, resulting in the denaturation of proteins or, in case of starch, the gelatinization of starch. Possible extrusion parameters can be sorted into three groups, namely process parameters (moisture content, screw configuration, die size, etc.), system parameters (energy input, residence time, etc.), and product characteristics (color, nutritional value, texture, flavor, etc.) (Chen et al. 2010).

In the following, the individual components of an extruder will be explained and described in detail. Figure 5 illustrates the components.

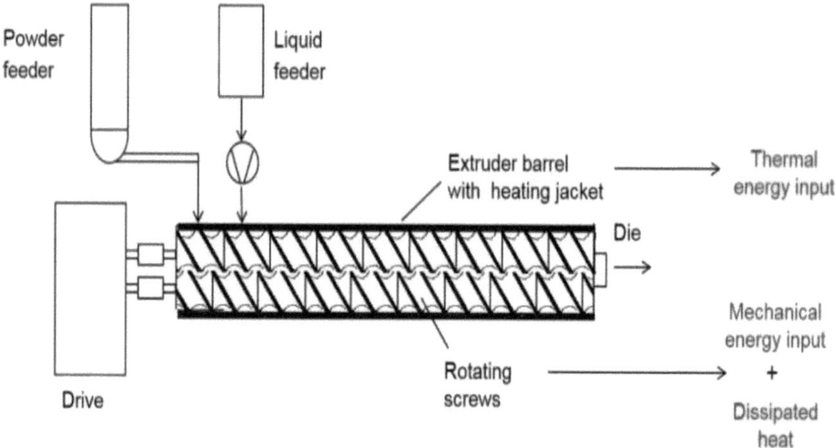

FIGURE 4: SCHEMATIC ILLUSTRATION OF A TWIN-SCREW EXTRUDER (WOLZ, 2017).

To be able to convey the highly viscous compound along the screws and to push it out of the extruder die, sufficient torque is required. This is provided by a powerful **electric drive** (Riaz 2000). Dry solid components are added as premix via a **volumetric feeder**, which ensures a uniform and continuous feed into the extruder, even if the bulk density of the raw materials fluctuates. Feeding can be executed either with one screw or with a twin-screw feeder, which is preferred especially for sticky compounds. The feed rate is adjusted in relation to the screw speed so that a constant mass flow rate is achieved. It is equally important to have a good injection pump for liquid ingredients, as even small variations in water quantity can have a significant impact on product quality. (Guy 2001) Moisture content is thus an important factor in system parameters and product characteristics. Higher moisture content results in lower viscosity of the dough in the extruder and shorter residence time (Chen et al. 2010). Additives such as sodium chloride, buffering agents, flavorings, colorants are added only after the extrusion process with fat as carrier to avoid losses (Belitz 2008). The **extruder barrel** tightly encloses the extruder screws and is therefore made of hard alloys to withstand the pressure that is build up. The inner surface of this barrel is grooved to prevent materials from slipping off the walls. The extruder housing is equipped with a heating jacket to control or maintain a constant temperature. (Guy 2001) Here, heat is also supplied to the compound by heat transfer, by dissipation of mechanical energy, and by introduction of steam that can be injected through the barrel wall (Harper 1981)

Screw extruders contain one or more **screws** that are co-rotating. The rotating screws generate high pressure, which can be adjusted with various elements to obtain effective mixing and the required product conformation. The number and design features of the screw elements depend on the process requirements. The rotation inside the barrel moves the material from the feed side to the discharge side. (Harper 1981) Generally, the screws are only suspended on the input side of the cylinder and rest on the product at the output end. Hence, the greatest load, and therefore the greatest wear, occurs at the discharge end (Guy 2001). Finally, in a High Moisture Extruder, the protein mixture passes through the **cooling die**. This cooling unit prevents the matrix from expanding and facilitates the formation of an anisotropic, or fibrous, structure. (Wittek et al. 2020) The die can consist of one or more openings through which the extrudate flows. The conformation of the openings gives the final product its shape. (Osen et al. 2015)

5. CONCLUSION AND FURTHER CONSIDERATIONS

With a growing shift in consumer thinking towards healthy and sustainable diets, the demand for vegan meat substitutes is growing. Typically, dried or fiber-based substitutes are made from soy protein by extrusion. The industry has an increasing interest in further exploiting the capabilities of the extruder to offer more diverse meat alternatives. For this purpose, a fundamental understanding of protein structures is elementary to be able to specify the process parameters for the final product. This also includes knowledge of the protein fractions of soy, as these influence the solubility of the protein with their different isoelectric points and must therefore be considered in basic research.

Soy protein is well suited for meat alternatives, as it is available worldwide at low prices and it is characterized by diverse functional properties and high nutritional value. It is available in three purity grades - isolate, concentrate and meal. The most commonly used is soy protein concentrate, which can be produced by either the acid-wash method or the alcohol-wash method. Protein concentrates from an alkaline wash have significantly better functional properties than those from an alcohol wash because alcohol has a strong denaturing power.

The extruder is used in many areas of the food industry. A distinction is made between low moisture extrusion (water content < 35%) and high moisture extrusion (water content > 50%). Low moisture extrusion produces porous expanded structures that must be rehydrated by the consumer before consumption. The second method produces fibrous, anisotropic structures that are intended to resemble animal muscle meat.

In both methods, a protein slurry is initially added to the extruder via a feeder, in which the mass is subjected to strong mechanical forces and high temperatures. This causes the macromolecules to lose their native structure and form new bonds. This occurs primarily in the area of the screws. The outline of the output gives the final products their desired shape. The only difference in the design of the extruder in both methods is the cooling die, which has to be attached at the end in the case of high moisture extrusion.

However, the mechanisms of protein-protein interactions during extrusion are still largely unknown. However, they play an important role in the final texturization. This lack of knowledge is the major limiting factor in realizing the full potential of this process for meat substitute production.

In future investigations of plant proteins regarding their suitability for extrusion, a special focus should be placed on the mixing behavior of biopolymers during thermomechanical treatment. Since a finished product could not only benefit from the nutritional advantages, but entirely new rheological and functional properties could be created.

6. REFERENCES

Belitz, Hans-Dieter (2008): Lebensmittelchemie. 1st ed. Berlin, Heidelberg: Springer Berlin Heidelberg.

Chen, Feng Liang; Wei, Yi Min; Zhang, Bo; Ojokoh, Anthony Okhonlaye (2010): System parameters and product properties response of soybean protein extruded at wide moisture range. In *Journal of Food Engineering* 96 (2), pp. 208–213.

Eldridge, Arthur C. (1982): Determination of isoflavones in soybean flours, protein concentrates, and isolates. In *Journal of Agricultural and Food Chemistry* 30 (2), pp. 353–355. DOI: 10.1021/jf00110a035.

Gomes Arêas, José Alfredo (1992): Extrusion of food proteins. In *Critical reviews in food science and nutrition* 32 (4), pp. 365–392.

Guy, Robin (2001): Extrusion cooking. Technologies and applications. Boca Raton Fla., Cambridge Eng.: CRC Press; Woodhead.

Harper, Judson M. (1981): Extrusion of Foods: Volume I. 1st ed. Boca Raton: CRC Press.

Khan, Rizwan Hasan; Siddiqi, Mhoammad Khursheed; Salahuddin, Parveen (2017): Protein Structure and Function. Aligarh Muslim University, India. Interdisciplinary Biotechnology Unit.

Kristiawan, M.; Micard, V.; Maladira, P.; Alchamieh, C.; Maigret, J-E; Réguerre, A-L et al. (2018): Multi-scale structural changes of starch and proteins during pea flour extrusion. In *Food research international (Ottawa, Ont.)* 108, pp. 203–215.

Liu, Ke Shun; Hsieh, Fu-Hung (2007): Protein–Protein Interactions in High Moisture-Extruded Meat Analogs and Heat-Induced Soy Protein Gels. In *J Amer Oil Chem Soc* 84 (8), pp. 741–748.

Mession, Jean-Luc; Chihi, Mohammed Lazhar; Sok, Nicolas; Saurel, Rémi (2015): Effect of globular pea proteins fractionation on their heat-induced aggregation and acid cold-set gelation. In *Food Hydrocolloids* 46, pp. 233–243.

Osen, Raffael; Toelstede, Simone; Eisner, Peter; Schweiggert-Weisz, Ute (2015): Effect of high moisture extrusion cooking on protein-protein interactions of pea (Pisum sativum L.) protein isolates. In *International Journal of Food Science & Technology* 50 (6), pp. 1390–1396.

Osen, Raffael; Toelstede, Simone; Wild, Florian; Eisner, Peter; Schweiggert-Weisz, Ute (2014): High moisture extrusion cooking of pea protein isolates: Raw material characteristics, extruder responses, and texture properties. In *Journal of Food Engineering* 127, pp. 67–74.

Riaz, Mian N. (2000): Extruders in Food Applications. 1st ed. Boca Raton: CRC Press.

Schuber, Sven (2008): Biochemie: UTB GmbH. Available online at https://scholar.google.de/scholar?hl=de&as_sdt=0%2C5&q=Sven+Schubert.+Bioche mie.&btnG=, checked on 6/4/2021.

Schuchmann, H. P.; Danner, T. (2000): Product Engineering Using the Example of Extruded Instant Powders. In *Chemical Engineering & Technology* 23 (4), pp. 303–308.

Shand, P. J.; Ya, H.; Pietrasik, Z.; Wanasundara, P.K.J.P.D. (2007): Physicochemical and textural properties of heat-induced pea protein isolate gels. In *Food Chemistry* 102 (4), pp. 1119–1130.

Statistisches Bundesamt (2020): Vegetarische und vegane Lebensmittel: Produktion steigt im 1. Quartal 2020 um 37%. Available online at https://www.destatis.de/DE/Presse/Pressemitteilungen/Zahl-der-Woche/2020/PD20_30_p002.html, updated on 7/21/2020, checked on 5/17/2021.

Vasudevan, D. M.; Sreekumari, S.; Vaidyanathan, Kannan (2011): Textbook of biochemistry for medical students. 6th ed. New Delhi: Jaypee Bros. Medical Publishers.

Vázquez-Rowe, Ian (2020): A fine kettle of fish: the fishing industry and environmental impacts. In *Current Opinion in Environmental Science & Health* 13, pp. 1–5.

Wang, H.; Johnson, L. A.; Wang, T. (2004): Preparation of soy protein concentrate and isolate from extruded-expelled soybean meals. In *J Amer Oil Chem Soc* 81 (7), pp. 713–717.

Wittek, Patrick; Zeiler, Nicole; Karbstein, Heike P.; Emin, M. Azad (2020): Analysis of the complex rheological properties of highly concentrated proteins with a closed cavity rheometer. In *Applied Rheology* 30 (1), pp. 64–76.

7. FIGURES

Figure 1: Different levels of protein structures (Vasudevan et al., 2011, p. 31) 3

Figure 2: Processes for the production of soy protein concentrate via alcohol-wash method(Wang et al. 2004) 7

Figure 3: Processes for the production of soy protein concentrate via acid-wash method (Wang et al. 2004) 8

Figure 4: Schematic illustration of a twin-screw extruder (Wolz, 2017). 11

Tables

Table 1: Protein Fractions in Soy Protein 5

YOUR KNOWLEDGE HAS VALUE

- We will publish your bachelor's and
  master's thesis, essays and papers

- Your own eBook and book -
  sold worldwide in all relevant shops

- Earn money with each sale

Upload your text at www.GRIN.com
and publish for free